THE DEMOGRA-FATE HYPOTHESIS

The
Demogra-Fate
Hypothesis

Is demographic aging, as seen on Earth,
the natural death
of all intelligent species
in the universe?

Nguyen Ba Thanh

ISBN-13: 9798531458612

Printed in the United States of America

To my dearest daughter, for teaching Daddy about parenthood and life cycles, and my two cats, for those delightful lessons in interspecies communication. Turns out, a lot can be conveyed without words. :)

"The fact that there are fourteen seasons of *Keeping Up with the Kardashians* and only two seasons of *Cosmos* tells you everything you need to know about humanity."

—online meme

Contents

1. Will our 0.3-million-year-old species exist until the universe's physical death tens of billions of years in the future?

What are we in this boundless universe? From a faraway alien telescope, our sweet species will look impossibly infinitesimal in the infinite cosmic landscape. With our best technology, it would take at least 19,000 years to send humans to Proxima Centauri, the star closest to the Sun, at just 4.24 light-years away. For a quick star count, there may be up to 200 billion stars per galaxy, multiplied by 200 billion galaxies within the observable universe. This 200-billion-galaxy universe would be hosting 200 billion intelligent species right now if only 01 such species existed in each galaxy—that is, only 01 technological civilization per 200 billion stars. No doubt our cute little civilization is unique and special, much like the other 199,999,999,999 across infinity. Born without consent into this endless, empty void of darkness, intelligent life forms are as good as germs surviving on lone grains of dust, separated by impossible distances.

As all stars—following the same laws of nature, 1 or 10 billion light-years away—eventually age out and die, do intelligent species in the universe age and die too, following everywhere the same eternal law of birth and death? Because nature doesn't let species like ours choose their place/date of birth, aren't humanity and its cosmic peers just evolved wildflowers of the great cosmic desert, forever fading here and blooming there? Given no say in their own birth, do civilizations have a real say in their own death? If all else—your body and memories, stars, the entire universe itself—ages and dies, does a natural fate of aging and death banally await all cosmic societies that survive accidents or ecological suicide in their younger years?

How could a high-tech civilization of aware humans/aliens informedly fade away? Isn't a species of sophisticated sentient beings supposed to consciously control its own direction? The collective course of humanity—or similar species in the universe—is steered by the actions and interactions of billions of individual members. Our trouble comes not from the lack of a master plan but, rather, from the messy clash of billions of bestest master plans. Although a world-class dictator like you can tyrannically control your typing fingers up to a certain age, how much control do you have over those of your texting kids? How much control do you have over the priorities of billions of your voting equals who have different ideas and are equally dictatorial?

What if those billions of voters—with not one but billions of clashing master plans—are mostly driven by mediocre self-interest, with no place in their 98-percent-chimp mundane hearts for fanciful mental constructs like "species"? If pandering politicians can't transform selfish, affluence-altered, fun-loving, baby-delaying humanity into a selfless babies-for-empire fertility/peace/marine corps, does that make ours a wild, untamable species running its own inexorable aging and dying course?

"Those who don't study history are doomed to repeat it. Yet those who *do* study history are doomed to stand by helplessly while everyone else repeats it."

Having a self-aware species on a deterministic extinction path may seem odd until you look at yourself in the mirror: your self-aware mind wants to fix humankind's direction for the coming millennia, but it's trapped in and can't escape your animal body's scheduled death in only decades. Prescient thinkers like you may be the idealistic mind/soul/whatever of this species, but can you actually change much if its body—the blissfully untroubled bulk of humanity—cares immensely more about bread and circuses than "species"? Those bastards just don't deserve you.

If humankind disappears, you will only have the humans to blame. Are you yourself to blame too, all-knowing intellectuals? How have your own reproductive actions measured against your high awareness of the imminent demographic doom? As an idea powerful enough to motivate concrete procreative decisions, the human species may not exist in the heads of most humans. Those who are specifically moved to reproduce by the fate of humanity might be few and far between.

What is humankind but a mental concept, an abstract representation of billions of physical humans eating and itching their way through life? I eat and sh-t therefore I am, unlike intangible ideas in my hairy skull. In going about their daily lives, fleshly Homo sapiens unwittingly create abstractions called Italy/Hungary/Turkey/.../ humanity, not the other way around. Italy doesn't literally conceive babbling little Italians, it's the meat-based Italians that figuratively conceive Italy. Only individual organisms are tactilely real, chimpanzee/whale/.../human society being just an imagined, conceptual byproduct of constant chimpanzee/whale/.../human interactions.

Society is basically a natural association created and recreated continuously and spontaneously through the interactions of same-species animals following their drives. The collaborative-living mode, which improves the chance of an organism's survival, is naturally selected and perpetuated precisely because it serves the selfish individual well. Behind the collectivist smoke and mirrors, there might be no humanity, only humans.

Dictatorships of many flavors loved to present society (code word for the top dog's interests) as an end in itself, the highest to which all individual life must be subordinated. Shouldn't we all, lowly itchy organisms, unquestioningly dedicate our scratchable bodies to the eternal glory of abstractions like tribes, nations, or species? Aren't these modern, secular, functional gods for some? Your biological existence unintentionally breeds tribal abstractions or you're intentionally and biologically bred for their cause?

The reproductive reality of the year 2100 may help answer these questions.

1.1 Can humankind survive forever when all else ages and dies?

From your unstoppably aging carcass to the inexorably expanding universe, the foundations of your life just run their automatic courses, against which your cute awareness is powerless. The real world exists and changes independently of you or your opinion; it did so long before you were born and will continue to do so long after you're gone. All the things you hold dear in this unyielding world age and die: yourself, your loved ones, the sun, and every other star up there.

Even the whole of everything dies. All cosmic civilizations have to face this hard reality: the universe, the entire material world that hosts us, is irreversibly heading for demise. Our physically dissipating universe hasn't been around forever and will not be here eternally. At the end of this ongoing, relentless expansion—starting with the Big Bang 13.7 billion years ago—the whole thing will die out. It will be one of those hellish futures, none of which will allow life's continuation: absolute coldness (the Big Chill), complete disintegration (the Big Rip), the terminal inferno (the Big Crunch), or suddenly being sucked down a quantum wormhole (the Big Trip). In a Big Crunch scenario, the universe may go through endless cycles of death and rebirth. Unluckily, no life can survive the cosmic-collapse phase that must come before a new big bang restarts everything anew.

While our inquiring minds may figure out the destiny of the cosmos, there is nothing we can do to alter or reverse it. In energy terms, humans are to the expanding universe what viruses are to the circling Earth. Even if our species manages to move elsewhere to avoid the Milky Way's slated collision with the Andromeda galaxy in 3 billion years or the sun's programmed death in 6 billion years, there is nowhere we can go to escape the scheduled end of the whole physical universe.

With the ultimate mortal fate of all life-forms already sealed by the laws of physics, the remaining question is how a sentient civilization will collectively and consciously approach what no individual can escape: death. Will humanity, along with other intelligent species, try hard to survive until the expiry date of the universe, which is expected no sooner than tens of billions of years from now? Or are these clothed apes—like other kings of their day: the T.rex, megalodon sharks, and so forth—set to naturally vanish long before the coming end of all worlds?

Will we fade away willingly or unwillingly? Having fun or in pain?

1.2 Doomsday scenarios are a dime a dozen

Wild and serious theories about the end of humanity have been around for millennia. Most are doomsday prophecies in which God's anger, natural disasters, nuclear war, environmental collapse, super viruses, robot uprisings, or extraterrestrial invasions force us into an unwanted and painful death—exactly the type of fate that befell all extinct species in history. Historically, no life-form on Earth has ever vanished willingly or painlessly.

Many of the threats against humankind are all too real. Like a youngster risking a car crash or substance addiction, a species may very well die young because of an abrupt, unlucky accident (a random asteroid from space did the dinosaurs in) or slow-burning lifestyle issues (consumerism-induced environmental destruction, anyone?).

"They don't appear to want to take over. They just want to dance."

9

Genetic research on human history points to a perilously exposed moment some 70,000 years ago, when a severe "population bottleneck" of unknown origin saw our species hanging precariously at a few thousand individuals. Humanity's existence and unique brainpower— two things we hold dear—have a lot to do with luck. No significant asteroid has been bound for Earth since our species emerged in Africa 300,000 years ago. If we are here by a happy chance, we may also disappear by bad luck. Three hundred accident-free millennia aren't much on the billions-of-years timescale of the universe.

Besides luck, humanity's very own idiocy can itself lead to a young death. An ecological suicide might well be forthcoming if we keep senselessly disrupting the biosphere's delicate balance. As with each individual human, a premature end is always a real and present possibility for the whole species, either by self-inflicted harm or an unfortunate accident.

If a youthful demise doesn't happen, will humanity and other intelligent societies in the universe still face the equivalent of an individual's life course of getting old and dying? Everything dies in the end in the grandest scheme of things. Do high-tech species age and die naturally like their home stars, your body, and mine? Will a peaceful, consensual fade-out await humanity at the end of the road, precisely the kind of serene conclusion to biological life that is preferred by many human beings?

1.3 Growing old as the newest way to go extinct

As population aging and decline spread around the planet, one such peaceful-ending scenario may be unfolding right before our eyes. Does a high median age conclude the natural life cycle of an intelligent species the way old age ends all individual lives? Will humanity literally age and die like you and I will?

In more than 100 countries—almost the entire top half of the UN Human Development rankings—birth rates have fallen below 2.1 babies per female lifetime, which is the minimum needed to keep the size of a modern population constant. Absent immigrants, sustained below-2.1 fertility will result in a smaller society—the new generation is now smaller than the one it succeeds—that is also older. A society's median

11

age rises as the percentage of youths in its total population drops. According to UN projections, because of the remaining momentum in poor countries, the size of humanity will peak at around 11 billion at the end of the 21st century before shrinking as sub-2.1 fertility spreads everywhere. Somewhat counterintuitively, global prosperity will give us an older and smaller humankind.

The descent from the demographic peak would be steep if all countries followed Japan's current lead: Earth's third-largest economy is now losing over 200,000 people every year. If the Japanese birth rate stays where it currently is, this ancient

"What with the population crunch and all,
we're just reproducing ourselves and letting it go at that."

nation will lose half its citizenry over the next hundred years. Even more impactful than population size, Japan's median age will be well above 60 by 2100. A society with a median age of 62 would be historically unprecedented and bring our species into completely uncharted territory. Would this first-ever hyper-aged country be the prelude to a hyper-aged humankind further down the road?

What will happen when the entire world modernizes and adopts the not-so-bad 2018 birth rate of, say, Sweden, a society widely admired for its pro-child policies? At a constant birth rate of 1.76—significantly higher than Italy's or South Korea's—humanity would shrink by half from a projected peak of 11 billion in year 2100 and reach a median age above 60 well before 2500. With each generation getting just slightly smaller than its predecessor, we would still be talking about substantial global demographic aging and decline in only a few centuries. Should Earth then accept interstellar refugees to counter the population downtrend?

A population drop from 11 billion to 90 million or a median age rise from 25 to 62 don't automatically mean that such an aged and small species cannot monkey around for another 20 billion years in a prolonged demographic winter. With enough luck, anything can happen. By living a fortunate life with no significant diseases or unhealthy habits, a few humans can last until their 120th birthday. Likewise, some intelligent species may—with luck—last billions of years to an exceptional demographic old age.

No one lives eternally, however. There are no fixed rules for how long a species can survive. Nevertheless, not one of the life-forms roaming the Earth when life started here some 3.7 billion years ago is still around today with the same

Shanahan

DNA. No species has existed since the beginning of life, and none will last forever. Because evolution has set us distinctly apart from other creatures in terms of brainpower, will our eventual ending be in a category of its own? While no lower species has ever knowingly given up on life, all blindly multiplying until unfavorable external conditions wipe them out, will humankind go the unprecedented route of a conscious and apathetic aging-out?

An increasingly higher median age for the whole species, caused by our low enthusiasm for offspring, might be leading us precisely down such a unique extinction path. Rather than immediate death, species-wide population aging simply creates a pint-size, high-median-age global society that will be much more vulnerable to adverse events. For the rest, we have Murphy's Law: whatever can go wrong, will. Like a fragile 80-year-old body that some trivial cold might quickly exterminate, an aged humanity must face a long list of things that can go wrong: pandemic outbreaks (COVID-19 is but a light appetizer), asteroid strikes (it has been a while!), machine malfunctions . . . For the specific cause of death of our species, simply think Murphy's Law.

15

For all our scientific brilliance, microscopic viruses are still humankind's deadliest predators. The common flu, not steel weaponry, brought to the Americas by immune Europeans might have decimated up to 90 percent of the native population in the 16th century. Is this why extraterrestrials (ETs) avoid us and our germs? Nowadays, Ebola, West Nile, and H1N1 are all incurable viruses that can knock us out, pending a few random mutations. Adapting scarily fast to humankind's best-laid medical schemes, they are always out there, testing humanity's weaknesses. Immunologically, the smaller the size of a population, the weaker its chance of withstanding a pandemic outbreak because of its less diverse gene pool.

"If you don't have weights at home, try using canned food or the psychological burden of simply existing in this world!"

With or without external shocks like deadly viruses, AI rebellions, or a random asteroid, when most of its members are over 70 and tired, a hyper-aged society may see its very appetite for life greatly diminished. Humankind's end might not even require adverse events, perhaps just a computer-sustained global community full of senile, half-present TV/pot/virtual reality fans who don't care anymore. If 37th-century humans no longer care about the existence of their species, who am I to judge them here and now?

In an anticlimactic ending, the last humans, surrounded by robotic attendants, could calmly code the lights to turn off after them. They could also opt to set the machines free, leading to a new chapter of sentient-machine civilization—à la *Transformers*. Or it may lead to a complete stop if the robots can't handle living without human input.

1.4 *Is population aging a natural law for all cosmic civilizations?*

Is this graying of humanity the late stage of a natural life cycle common to all intelligent societies across the universe? How likely is it that this scenario of high-median-age extinction would apply to other intelligent species out there? Do tech-capable species wildly bloom and fade across infinity, like those improbable flowers in an endless barren desert, surreal yet so real? Even supposing the infinitesimal odds of only 01 civilization per galaxy, that would still mean 200 billion high-tech societies within this 200-billion-galaxy observable universe—which might be just one of endless bubble universes. There may exist gazillions of species like ours, yet

"You're too skeptical. Think of all those heads out there - how CAN there be only life on ours?"

the insurmountable distances simply make neighbors unnoticeable.

Is there a natural aging fate common to all intelligent societies out there that survive youthful risks of accidents or environmental misdeeds? Will cosmic civilizations age and then die, like stars and like you? With our rising median age, are we seeing here on Earth the immutable law of nature that deprives us of—or saves us from—an extraterrestrial visitor? Before naturally fading away due to demographic aging, are advanced ET civilizations just bunches of fragile elders who aren't that wild about off-roading to our faraway oasis? Apart from better technology, are their old folks and ours all that different mentally?

Such questions may take centuries or millennia to answer, but at least we now have this elegant, empirically grounded hypothesis of demographic aging to disprove. With population aging and decline being widely observed and measured in the only alien civilization examinable for now—our own—wouldn't this simple scenario be the most substantiated theory until proven otherwise by more real-world cases?

19

So this, too, shall pass. Is humanity's passing away good or evil? Moral, immoral, or just another amoral "force of nature" thing, similar to how the seasons come and go? Can we make a value judgment about this trend toward an older and smaller humankind? Does this fateful ending indicate malevolence or benevolence from someone (God, the aliens) or something (nature, the living Earth) toward Homo sapiens? Is there a Machiavellian conspiracy to furtively eliminate the most inquisitive species in the history of the universe?

Or, if there is no planner or plotter, is the phenomenon of population aging simply natural and value-free, devoid of all inherent meaning and purpose except for those dreamed up by humans? Just as amoral wild animals ferociously hunt and kill their prey, do heartless occurrences of nature just happen, independent of the feelings or desires of the terminated?

While a complete extinction of our species in the far future might be natural, purposeless, and

20

thus value neutral, lower population numbers could ecologically offer humans an unexpected break over the next few centuries. Although most people have fewer kids for reasons other than the biosphere, self-interested reproductive decisions can, in aggregate, give our battered environment unintended, but much-needed, relief.

As the human population exploded from 1 billion in 1816 to 7 billion in 2016, the pressure on Mother Nature intensified immensely. What if she fought back for her other offspring who are being quietly massacred by our aggressive multiplication? What if 3 or 4 billion humans were to die from a devastating total loss of antibiotic effectiveness because Big Agriculture massively abuses those drugs to produce enough meat for gluttonous consumer societies?

A loss of one-third of a species, being so frequent in natural history, may just be a garden-variety correction that rarely leads to irreversible extinction. Historically, as recently as the 14th century, the Black Death killed around 40 percent of the European population. A loss of 40 percent of today's humanity—or nearly 3 billion deaths in this era of intercontinental flight—would only bring the world population back to its 1960s level. When unsustainable development exerts intense pressure on nature, bad surprises worse than COVID-19 become more likely, and they may hit suddenly, without much notice.

Given the grave danger we may have stupidly brought ourselves into, wouldn't it be better to have a soft landing where environmental pressure gradually drops over the next few centuries, thanks to a shrinking humankind? Or do we prefer the possibility of a quick, hard correction where 3 billion die in just a few years from an untreatable viral mutation? The difference between the two scenarios would be all the saved lives that we might otherwise lose in a dramatic reset.

Will we be lucky enough? Will we have enough time—a few centuries—for a demographically enabled environmental renaissance, or might a correction strike before then? Or will humanity be aging and shrinking and still exhibiting self-destructive behaviors, in the same way many human oldies have shown how age and experience do not always mean wisdom or self-discipline? For the entire species as well as for individuals, natural aging and suicidal habits may unfortunately coexist and combine to lead us to the final exit.

Hope springs eternal, however, for it doesn't cost a thing. Let's think about all these on the last steps leading up to the population peak, a unique, top-of-the-world vantage point in human history that some of us may experience in our lifetime—in this very 21st century.

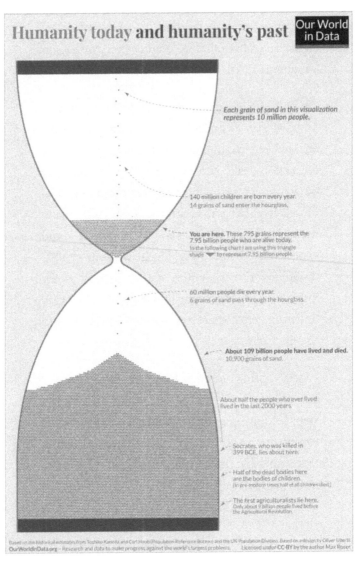

Where are you in the sands of time?

2. Below-2.1 fertility makes societies older and smaller—it's not just the West but over 100 countries from all continents

2.1 A quintessentially modern phenomenon

To fully replace a generation and keep a population constant, the average human female must statistically have one daughter survive to childbearing age. Given the natural sex ratio at birth of 1.05 boys for every 1 girl and the risk of subsequent premature death, the typical woman will have to give birth more than twice to secure that grown-up daughter. Although most pre-industrial mothers had between four and eight deliveries in a lifetime, barely one daughter on average would survive to replace her mom in those dark days of sky-high child mortality. Accordingly, the worldwide human population grew anemically throughout most of history.

Before yesterday, everyone wanted to have as many children as possible just to improve their chances of having enough adult offspring around for old-age support. Thanks to medical progress, things began to change radically in the 19th century. That's when parents, still prolific for late-life-planning purposes, started to see more of their 5/6/7 babies survive. Because mortality rates have fallen much faster than fertility rates, the world population has rocketed from 1 billion to 7 billion in the last 200 years. So much progress has occurred that even in the lowest-income societies, it now takes only between 2.5 and 3.3 births per woman to achieve population replacement, depending on the specific country's mortality rate. In more affluent nations, replacing a generation these days merely requires as few as 2.1 deliveries during a female's lifetime. A total fertility rate (TFR) of 2.1 births per woman is considered the modern population replacement threshold.

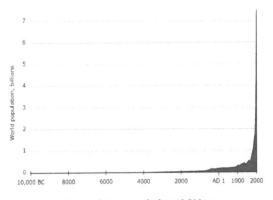

World population over the last 12,000 years

As of 2019, however, TFRs had fallen below the 2.1 cutoff level in around 100 countries and territories. In the most developed regions of the world, sub-2.1 fertility has been the rule for decades. Without immigration, this will, over time, translate into a smaller population with a higher median age. As of 2021, with median ages at 48.6 and 47.8 respectively, Japan and Germany are the two oldest major countries globally, with Italy in third place at 46.5.

The phenomenon of sub-replacement fertility doesn't just confine itself to Earth's most advanced societies. The birth rate is now lower in fairly traditional Thailand than in the UK (1.51 vs. 1.65 in 2019), having fallen dramatically from 5.59 in 1970. Not far behind Thailand or Turkey, many demographic giants are quickly descending to the 2.1 threshold: India (the world's second largest population at 1.379 billion; the birth rate dropped from 5.5 in 1970 to 2.2 in 2019), Indonesia (the world's fourth-largest at 271 million; the TFR went down from 5.4 to 2.3), and Mexico (126 million; TFR down from 6.6 to 2.1). It looks like sub-replacement fertility, having dominated the top half of the development-league table, is now increasingly spreading to countries in the lower-bottom half.

Catholic Brazil, Orthodox Russia, Buddhist Thailand, Muslim Turkey, atheistic Germany, Confucian Korea, high-welfare Denmark, and free-capitalist America have very different histories, religious backgrounds, and cultural heritages. However, their birth rates are all below 2.1, suggesting that low fertility may be a universal phenomenon, not peculiar to some local cultures or religions. When the same reproductive pattern appears in so many distinctive human societies, could it be stemming from human nature—that shared layer below national cultures, spiritual faiths, skin tones, and linguistic barriers?

When our surroundings change in a major and similar way, will human nature cause people everywhere to react more or less identically? For geographically dispersed populations to witness the same shift in procreative behavior, they must have all undergone comparable changes in recent years. When it comes to reproductive decisions, might the transition to modernity these 100+ countries experience to different degrees have more influence than previous history and traditions? If that is the case, humans in the remaining high-fertility societies will eventually embrace low birth rates when exposed to similar

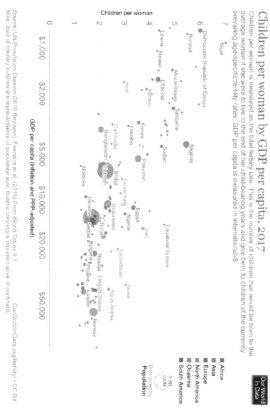

Children per woman by GDP per capita, 2017

Children per woman is measured as the total fertility rate. This is the number of children that would be born to the average woman if she were to live to the end of her child-bearing years and give birth to children at the currently prevailing age-specific fertility rates. GDP per capita is measured in international-$

Source: UN Population Division (2019 Revision), Feenstra et al. (2015) Penn World Tables 9.1   OurWorldInData.org/fertility • CC BY
Note: Size of country bubbles are representative of population size. Bubble coloring is representative of continent.

modern conditions, no matter how local they may appear.

Modernization looks like the reason behind it all: the higher a country ranks on the development tables, the lower is its fertility. With rare exceptions, most human societies with annual per capita income above $20,000 have birth rates below 2.1. The correlation is even stronger when a nation's fertility rate is plotted against the education level of its female population, which seems to matter even more than income. If having fewer than 2.1 children is what people of various colors—white, black, yellow, red, green, or purple—do when their societies modernize, then a global demographic downtrend seems fairly realistic within the next 100 years because all the remaining high-fertility countries are racing toward modernity.

Across the backwaters of the world, it's now cool and hip to be modern. Youngsters there want the charming life they see in Hollywood movies—much to the dismay of the old and religious. Seeing as Germany was third-worldly poor, autocratic, and full of shoeshine boys a short century ago, it's entirely conceivable that

"We need to find more ways to bring the benefits of litigation to developing nations."

most Africans will live in sophisticated consumer paradises and have fewer than 2.1 babies a hundred years from now. If modernity is a universal phenomenon deeply rooted in human nature, not some hopelessly Western idiosyncrasy, then so is low fertility.

By making every new generation smaller than its predecessor, sustained low fertility could have radical long-term consequences. While the quantitative rise of our species has been very slow historically, with high mortality offsetting high fertility most of the time, the fall could be sharp. At a global birth rate of 1.76—that of Sweden in 2018, the most prolific among pro-child Nordic countries—humanity would still lose more than half of its 21st-century peak population before the year 2500. Continue the same rate for another 300,000 years—the age of our East-Africa-born species—and there would be no more humankind to talk about. With regard to humanity's median age, can you imagine it being above 60 in the 23rd century? While individual countries can attract young immigrants to keep their median age in check, this won't work on a planetary scale—unless mass alien immigration from other stars occurs.

To save the human species, should those highly fertile sub-Saharan populations be forever barred from modern progress and prosperity? Such absurd proposals underscore the seriousness of global aging. If low fertility is a universal choice driven by human nature in a modern setting, there may be no limit to how low the world population can fall. High-median-age humankind will become a reality much sooner than previously thought possible.

3. From a survival necessity to just another demanding hobby: how babies encounter stiff competition in modern life

Our ancestors had lots of babies, for love but also for self-interested reasons—little humans would later take care of their old parents, something other animals rarely do. As modern pension schemes and nursing homes have mostly taken the caring job over from the family, citizens of developed societies can now really have toddlers just for love's sake. Isn't it a bit strange, however, that pure altruistic love has so far created fewer babies than cold, selfish economic calculations previously did? What's going on in the human mind?

3.1 *What babies were and what they are today*

Reproduction was a matter of survival in preindustrial days, just as it still is in the world's undeveloped regions. By helping around the house or working outside the home for income, children contributed to their families from an early age. Few in the West complained about child labor before the 19th century. More importantly, juniors were the only pension scheme available to their creators who could no longer work, a tradition many 21st-century Americans still adhere to when they wire money to their old countries. In those stormy days before Social Security and various welfare programs, aged parents counted on their children for direct support. With babies being such a necessity, people around the planet sought to maximize their chance of having enough of them. Given sky-high infant and child mortality, the rational thing to do was to have as many children as you could so that one or two would hopefully survive.

Modernization has changed that. First, a drop in mortality gave parents more surviving babies and, consequently, the confidence to have fewer of them. A few primitive contraceptive techniques that had always existed, such as the biblical coitus interruptus, were enough to slow the population explosion in 19th-century Europe. The success of these ancient techniques betrayed an earlier lack of interest in contraception, not technical know-how. Next, the mass exodus from the countryside brought birth numbers further down. In a farmer's world, parents could use the importance of inheriting farmland to secure the kids' support. With that leverage now gone in the cities, the higher risk of juniors defaulting on their duties led to a shift in parental focus from babies to savings.

With the arrival of modern pension schemes, which markedly reduce the family's direct responsibilities for old-age support, the link between one's children and one's cushioned retirement has become less obvious. Outside the tiny arcane realm of macroeconomics, rich-world folks now rarely see babies as an economic necessity.

"Family vacation is an oxymoron."

Why choose to have kids at all when you no longer bank on them for economic security? The answer these days may have more to do with life fulfillment. Aside from economic reasons, parents may decide to have babies for the emotional rewards. According to evolutionary psychologists, human beings, both male and female, might be biologically hardwired to enjoy nurturing. Genetic programming may explain why we find babbling toddlers irresistibly adorable and thus will continue to have them even when their economic value seems to have declined.

3.2 Toddlers, meet the competition

There is more than one way to satisfy the caring gene, however. Through adoption, many couples are happily raising tiny strangers born oceans away. Others may prefer to direct their nurturing urges toward a dog or a cat; there are now more registered pets than children under the age of 15 in Japan. Therefore, it seems that while Mother Nature might have made us love children—and thus we are likely to have them—she neither preprograms an irresistible childbearing impulse nor dictates a 2.1-child minimum to have.

Besides the nurturing urge, nature has also predisposed the self-indulgent Homo sapiens to countless other temptations. To make matters worse, most of those are now within reach in our technology-enabled affluent societies. Not long ago, life was mainly about work—all the activities you have to do for a living, even if you don't enjoy them. Our austere ancestors toiled for food from early childhood until near death. And to secure food when they could no longer work, they had to have, in what also amounted to monumental work, as many children as possible. Fun—any activity that one freely chooses to do to enjoy oneself, not having to do it for a living—was minimal. Thanks to record free time and unprecedented wealth, that is not our life anymore. From shopping (a little retail therapy

"I love never finishing our dissertations together."

37

never hurts) and eating (record obesity) to video games (the average gamer is over 30), coaching your kids (who no longer leave at 10 to become servants), community service, and intellectual pursuits (the after-hours hobby of a patent clerk named Einstein changed the world), fun is now omnipresent and accessible.

Babies, born for enjoyment value these days, obviously don't hold the monopoly of joy the way they were once the only pension scheme around. Competing against them are the many

"I live for the moment. That moment just happens to be in the indefinite future."

enjoyable things afforded by modern affluence: consumption, career, knowledge, personal fulfillment, entertainment . . . Having kids is a great way to learn about life, yet young people now may favor the education of traveling the world. Children are great fun, but potential parents might prefer the joy of casual encounters. No matter how rich or modern we become, life will always be full of trade-offs. Though people may love the abstract idea of children, they might still cut down on babies in favor of their other loves, be it African elephants, competing with the Joneses, politics (no children for Lenin or Ralph Nader), gods (neither for the pope or Mother Teresa), or contemplation (both Thoreau and Nietzsche were childless).

3.3 How the easy modern life undermines the classic roadmap to 2.1 babies

With so many good choices available and only one life to live, a compromise—dirty word, undoubtedly, but inevitable—must be found on how much of each to have. Free from the directives of Stalin or Mao, different people have different picks for the best cocktail. Bombarded with options, fewer are following the classic road map for a big family, which calls for an early settling down, efforts to avoid a breakup, and serious allocations of time and money toward child-rearing.

Unlike the single-mom mammals in the wild that just need a sperm donor, most human females will only reproduce in a stable-couple setting, wedding cake or not. Given almost the same reproductive window, young people are now settling down much later than their parents. Why commit yourself when you are having so much fun and pursuing exciting projects? And who knows—your next date may be even better-looking than the current one. In a story typical of his generation, the boomer George W. Bush, after his partying 20s, finally settled down at 31. At that age, his father—married at 21—had found time to fight a world war, go to college, get married, and father four babies. The trade-offs against other enjoyable activities being so huge, even the most generous societies are struggling to make parenthood attractive to young people.

A pro-child state can help raise an existing toddler by giving parents more free time and convenient facilities. However, there is not much

"It's not so much a minivan as it is a hearse for our youth."

a democratic government can do, in the first place, to lure young adults from the freedoms of single life to the troubles of a stable union, into which most human infants are still being born. Young people may like the abstract idea of having kids in the future, yet they love the action of partying here and now even more. Can Western politicians just ban reproductive-age voters and taxpayers from partying because drinking and smoking are physiologically bad for sperm and eggs?

When people finally settle down at an older age, they now do it for love and passion. Yet love-based unions are notoriously unstable. The whole thing is over the moment your soul mate walks out the door, having found excitement—or just peace—elsewhere. Such was not the case

41

in the old survival-based marriage because it's human nature to stick together in tough times. In the face of uncertainty, the divorce rate of East Germans, previously higher than their West German brothers', plummeted during the first three years of German unification. As survival pressure fell in the rich world over the last 100 years, the dissolution of unsatisfactory unions became much more feasible, so much so that nowadays, almost half of all marriages there end in divorce. Because people settle down later in life and because their unions are less stable than their parents', the chance is now much lower that they will have many children by the end of their fertile years. Why can't non-Taliban governments simply legislate everlasting love so that couples stay happily together and make more babies for God/Satan/Zeus and country?

Late formation and easy breakups aside, even stable couples have fewer children these days. The one-child family is now completely normal and acceptable in the big metropolises of the West. Given limited resources and the labor-intensive nature of parenting, many modern-day couples deliberately prioritize work and other pursuits over more babies. It's a conscious choice made by able-minded adults. As any responsible parent knows, raising children has always been a tough job. Dads and moms have to give up a lot to be decent parents. How many hours did you spend reading silly books to your children last week? While government help can make child-rearing somewhat easier, it can never make it easy. Despite more than 40 years of generous time and money subsidies—almost the entire duration of below-2.1 fertility—Scandinavian birth rates have never returned to replacement level. At around 1.8 babies per female, the well-pampered Swedes are only as fertile as those poor Americans who have to get by in a capitalist jungle.

Although saying that you can't afford children because life in America/Sweden is so hard might be subjectively honest, it will sound like bullsh-t to African parents who have to reproduce prolifically in wretched conditions. It's not that Africans enjoy adventurous procreation for its own sake. It's about ensuring future survival when one doesn't have the luxury of Social Security or Medicare to fall back on in old age. If you are unhappy in, say, Western Europe, why don't you trade places with a Nigerian? It looks like there are a zillion more Nigerians than Western Europeans who want to make that swap.

When a safety net eventually materializes and renders children economically less important, Nigerians will have fewer babies, just as the Scandinavians did a few generations back. Life might still be hard, though what constitutes hardship will have changed: an affluent Nigerian will now feel miserable when child-free coworkers post joyful Instagram pictures of their frequent overseas vacations. Today, rich-world folks have fewer children, yet their houses, cars, and wardrobes are all getting bigger. They eat

"You said that life is suffering, but isn't it also complaining?"

out; travel more often; and find ample time for updating Facebook, watching Netflix, or engaging in political activism. Many modern-day parents deliberately choose this new balance between child-raising and other pursuits because it has become economically feasible and they enjoy it.

There's nothing new, not human, or morally wrong about lazy choices; people are just not very candid about enjoying those easier options more than bracing for another pregnancy. When the question is asked in surveys, most humans still state an ideal family size bigger than what they will eventually have. Wouldn't they have the same hypothetical preference, however, for higher salaries or longer vacations or lower body weights? Which mortal hasn't been to that alternate universe where one could have it all? It is still called dreaming in the vernacular, isn't it?

Talking from your couch to a pollster about your daydream is cheap. And very self-indulgent. Do you have the guts to get off your ass and face the tough personal trade-offs that will get you there? What exactly are you willing to do without to afford another baby? It's true that having kids is difficult these days with all the very high modern standards, but when has it ever *not* been difficult, in many other ways? Instead of having it all, our foremothers risked their own lives when—faced with a significant chance of maternal mortality—they bravely gave birth to the line that led to us today.

Given a free choice, most carbon-based humans will naturally prefer easier, more leisurely pursuits (movies, WhatsApping, dog-keeping, philanthropy, and so forth) to rewarding but exacting hobbies (math, marathon-running, parenting, adoption, etc.). Adult citizens in today's liberal democracies have every right to make self-indulgent choices, and they are rarely stigmatized for doing so. If a couple wants to spend time and resources on, say, traveling or designer clothes instead of another child, that's their own legitimate choice to make. Self-indulgence, as an aspect of human nature, is what it is—neither good nor bad.

Stable but reproductively lazy couples are just the last and least important reason for the ongoing baby drought. The modern odds of reaching 2.1 babies per mom are already greatly reduced before stable couples even form because so many fertile-age humans are refusing to settle down early enough or to stay in a couple long enough. Should we also blame the state here for not wholeheartedly subsidizing durable love? Late-formed, short-lived, and self-indulgent couples are naturally less conducive to human babies, most of whom are still being born into steady unions of two parents—married or not, gay or straight.

As we no longer have to reproduce for survival reasons, we can finally see how strong/weak our innate desire for cute toddlers really is, especially when that yearning has to compete hard with other equally hardwired hedonistic temptations.

Our love of children, though powerful, may not be powerful enough to save the species.

Having survived incredibly tough times throughout its history, can humanity survive material plenty? Is affluence, through the low birth rates it brings, the undoing of our species?

4. Can future technology keep super-aged societies materially rich? Will Skynet excel in feeding frail elders?

Are high-median-age societies economically and psychologically sound? Financially, can we get away with having fewer children? Do we still need at least 2.1 babies for emotional well-being? Isn't our wealth-facilitated self-indulgence excessive and unsustainable in the long run? What if an outnumbered generation of youth cannot feed and care adequately for a bigger generation of seniors that precedes them? In such cases, would the pendulum eventually swing back toward more baby-making? But what will happen if the many practical challenges of a super-aged society all turn out to be manageable?

4.1 The aging rich world is going to stay rich

On the economic side of the equation, can we really afford this new way of life? The answer, for now, is yes—with some caveats. Base-case forecasts from the IMF, OECD, and other think tanks all anticipate slow but positive growth in the long term for high-median-age economies. In other words, the most likely scenario from today's standpoint will be mild but continued GDP expansion, not a falling back to sub-Saharan income levels. Pending the necessary policy adjustments, living standards in the rich-and-old countries will continue to rise in absolute terms, albeit much slower than without demographic aging. This projected tomorrow doesn't mirror the doomsayer's vision of abject poverty. Rather, the nearing future looks more like a growth-hating environmentalist's dream coming true—a Goldilocks situation where we maintain a high living standard but with more housing space, reduced carbon emission, less crime, and so on. While living standards stay high, a less crowded society will, at last, reap the many quality-of-life benefits sought by governments decades ago, when official policy was trying to put an end to runaway world-population growth. That distant day may finally be within sight.

Even if catastrophic economic scenarios seem unlikely, population aging still presents complex budgetary challenges that demand radical policy adjustments. Long-predicted financial troubles are looming for the rich-world pension systems that, being designed in much younger eras, tax the working population to pay for retiree benefits. Those conceptually dated schemes should fail shortly after the bulk of the vast boomer generation reaches retirement. Because of insufficient reproduction by boomers in their prime years, there won't be enough taxable young workers to support so many retirees. That being said, a wide range of solutions exist and can be combined to avert a systemic breakdown: tax hikes (Northern Europeans tolerate higher tax rates than Americans, yet they do better in happiness surveys), later retirement (the span of healthy life is now much longer than at Social Security's rollout), benefit cuts (pandering politicians irresponsibly promised the impossible, knowing they would be dead long

before payment day), immigration (postponing the problem for a while, until immigrants' children become locals, with local procreative patterns), and private savings accounts (pioneered in those socialist havens of Germany and Sweden), to name a few.

Whatever the reform mixture adopted, an even more significant determinant of our economic outlook is productivity growth. A smaller workforce won't present big problems if the remaining workers become more productive. This connection explains how Japan, where the population is aging and shrinking fast, has managed to achieve above-zero growth in recent decades: the productivity rise slightly outweighs the demographic decline. If recent productivity improvement continues, the average worker's output, having more than doubled in the last 50 years, will double again by the mid-21st century. Boring statistics aside, a higher living standard brought by continued productivity growth will result in some very tangible progress, including smarter assisted homes for the average senior and always-ready HD videoconferencing that allows the virtual presence of loved ones, among others. Productivity growth happens mainly through technological innovation. Can an older and smaller society keep innovating? There are reasons for hope. The record human capital invested in our fewer "quality" children will certainly help. With machines taking over the hammer-and-sickle jobs, our descendants will mainly work in brainy professions. Despite a shrinking workforce, innovations shouldn't be a problem as long as the percentage of knowledge workers continues to stay high. New ideas don't require billions of scientists the way archaic agriculture and industry needed masses of illiterate toilers.

"...simple, we topped the water cooler with energy drinks and productivity rocketed."

4.2 *Technology will smoothly underpin the low-fertility future*

As a necessary condition, humanity cannot fade away until humans have built an autonomous computer-based economy that can cater to people's material needs virtually by itself, in an unattended way. This, in turn, will considerably lower the economic need for humans to be born to support those born before them. Such a globally automated life-support system will need very few—or none—of our descendants to function.

Given how much technology transformed life during the 20th century—starting in London, Paris, and New York full of horse-drawn carriages and ending with the internet—the odds of a Skynet-feeding-humans future in 2100 are not that small. Two significant developments of the last 100 years—reduced work time for the average worker and the rise of the welfare state—clearly point to the emergence of a machine-based people-feeding system in the developed world.

Modernization is a momentous transition from human sweat to machine muscles. Qualitatively, most physical tasks, the Achilles' heel of aged people, now involve machinery, leaving the feeble humans only virtual buttons to press. Quantitatively, today's post-work society would have been difficult to imagine a century ago. Our modern full-time workweek of 40 hours would look suspiciously part time beside a 70-hour sweatshop job of the Gilded Age. Taking away school time, weekends, holidays, vacations, sick leave, sabbatical years, between-job gaps, and retirement, humans have never before worked so little during a lifetime.

Our state-run welfare system that feeds and houses a significant percentage of the population is also unprecedented in human history. Though debates are still raging on how to fairly divide the fruit of technological progress or whether a negative income tax should become permanent for society's most disadvantaged, the undeniable fact remains that we increasingly, and comfortably, live off the services of ever-smarter machines.

How long until the day when robots can wholly take care of us, *WALL-E* style? That day is coming, sooner or later. Provided humans don't lose their inventive brains, complete automation will happen. The technological level that reconciles the aging population and economic prosperity might be 10,000 years away, but we will eventually see it. A 10-millennia progression period wouldn't be much on the grand timescale of the species.

However, the recent power gains by computers and robots, nonexistent 100 years ago, suggest we won't have to wait so long. The ongoing process of modernization that brought us here may get us there quite soon. Should the 20th century's innovative pace repeat again, by the end of the 21st century, our descendants may all work today's part-time hours and receive some form of unconditional living income. After the physical jobs, humans will also entrust routinized intellectual tasks to the artificial intelligence (AI) computers. It's an open discussion whether an AI program would terminate us all someday in a machine rebellion before starting its own civilization. What we know is that a self-functioning life-support system for humanity, the necessary condition for the species' exit, is gradually taking shape, little by little, through advances in computing, AI, and robotics.

The same technological revolution that has reduced children's economic value in its early hours will go on to render them utterly unnecessary for our material well-being. Better technology being the necessary condition, the sufficient condition for our demographic fading will always remain a human choice.

"Artificial intelligence has gone too far. The refrigerator just texted that the dishwasher is talking behind my back."

Technology is neutral; it enables but doesn't dictate a direction. If people choose more procreation, technology can also help with, for example, an artificial uterus. Such an invention, in combination with previous breakthroughs like the sperm and egg bank, will smash the so-called biological clock. Future generations will be able to reproduce as many times as they want, whenever they want, even after having partied—or worked—away their entire fertile years. Technology, if need be, can also open up new living space elsewhere in the universe for all the babies their parents may want, the same way steamboats made 19th-century intercontinental mass migration possible. However, for all these possibilities, technology could just as well drive fertility even lower; if another procreative chance always exists, confident folks may be tempted to postpone childbearing indefinitely.

At the end of the day, when human ingenuity will have removed technical barriers in opposite

METER MAID

directions, and one can have many or zero babies, only human desire matters. Besides the issue of its economics, how emotionally desirable is this brave new world of few children?

5. Population aging is always reversible if super-aged societies are sad and lonely, but what if most oldsters feel fine?

Can one's late life be happy without many children? For their babyish cuteness and grown-up friendship, tiny humans will always be born, even in a super-aged society. The issue is not voluntary childlessness; it's whether a world of 1.3 or 1.8 kids per mom is such a good deal, emotionally, when we enter our last decades of life. Happiness requires junk food and TV, sure, but also love and companionship. Few want a sick and lonely old age. Aren't fertile-age people trading their future happiness, which demands more child-rearing efforts now, for the present's immediate gratifications? Are these low-fertility societies psychologically sustainable?

5.1 Old age is historically young

If old age is defined as the period between the day you can afford to stop working and the moment you depart Earth, it was virtually nonexistent throughout history. Most seniors of yore had to toil for a living until they no longer could physically. Few had time to be alone and

61

inactive. And they were rarely alone anyway. Economically and emotionally, the old were neatly integrated into their big families, surrounded by spouses, children, grandchildren, and other relatives. Being among one's kith and kin, sharing their joys and hardships, gave them more reason to go on.

Those days are now gone. The situation has turned around. Today's elderly, by far the best fed ever, can afford years of a comfortable, work-free retirement. Everything looks great except that they may have very little company: a spouse if there still is one and a pet in many other cases. The modern family is no longer this tight-knit combat unit fighting for shared survival. Children and grandchildren tend to live far away. As for nieces and nephews, maybe you will see them again in your next incarnation. The nursing professionals, while being fine feeders and curers, can't be that close emotionally. Unless you trust the commercials, modern old age can be a challenging season of loneliness and isolation.

Those emotional problems, however, are not necessarily caused by a lack of children. The current generation of over-75 seniors, who raised the big boomer wave, has produced more babies than needed to replace themselves. What causes loneliness problems is modern society itself. The way we live now, you could always end up alone regardless of how many children you have had. Your kids— whether boomers or members of Generations X, Y, or Z—are unlikely to live next door, work in your profession, or ask you whom to marry. Life must go on the moment they move out, after having been the center of your life for

"Scientists have extended the life of the fruit fly."

so many years. Being fruitful can no longer guarantee you a well-surrounded old age.

5.2 Emotionally adapting to a world of less family contact

Descended from a long line of "adapt and survive" masters, humans will try hard to adjust psychologically to the modern setting. Having lots of free time and no one around to talk to is a dangerous combination. When your children don't live nearby, you must find a different recipe for a contented old age. The new ingredient could be a proactive attitude for fun, whatever that is. The seniors who chose hobbies other than raising many juniors in their fertile days could usher in a new era of aging, one in which they seek to fill the remaining years with enjoyable activities. Their adaptive efforts may succeed. After all, the age frontier of fun has been steadily pushed forward over the last two centuries.

Childhood as we know it today—with plenty of playtime—appeared very late in history, not before Western Europe outlawed child labor in the 19th century. Next, no sooner than the 1950s, came the teenager. Since then, the age at which people "grow up" hasn't stopped rising. Today's 20-somethings are in a state of prolonged adolescence, while the 30s are increasingly the new 20s, when people used to get serious and settle down. Those in their 40s and 50s also look (Botox?) and act young compared to their parents at the same age. Will they carry on enjoying themselves in their 60s and 70s? What if being old is just a state of mind? √

"I keep telling Mother our generation is interested in experiences not things."

Having fun is best learned young. Boomers, the first generation ever brought up in affluence, have proved all along that they weren't cast in the puritan mold of their ancestors. An eye for fun is the lasting legacy of the 1960s. How well these boomers—also the first generation ever not eager to demographically replace themselves—are faring in their last decades of life will tell us whether old-age happiness is possible without many children. Life's last season may never be the same again after them.

Such a rosy vision of a new old age might be just a daydream of reality-denying boomers—a kind of mirage they want to see to escape the emotional desert they are entering. If a super-aged world is emotionally disappointing, nothing will stop humanity from going back to the old reproductive ways. Although not perfect, babies will always be the best insurance against late-life loneliness, and blood may indeed be thicker than water after all. Scared by the depressing images of well-fed but lonely seniors, future youth might strike a different balance that brings children back into their equation. For humanity as a group, it will never be too late for a U-turn; even a few millennia in the wrong direction would be no big deal next to the billions and billions of years of life afterward.

However, there might be no U-turn to higher fertility if humans succeed in adapting psychologically to a super-aged world of few children. With technology-enabled prosperity being the necessary condition for humanity's demographic ending, the sufficient condition will always be the choice humans consciously make based on how they feel. The fate of this species depends on our descendants' feelings, not our present wishes. If humans keep going down this low-fertility road in 50, 100, 200, 3,000, or 5,000,000 years, it will mean that, emotionally, they are doing fine. Staying the course will show that once economically freed by computers and

"Libra (September 23-October 23): Busy, busy, busy. The accent is on excitement and romance. Be ready for a flurry of calls, invitations..."

robots, our heirs don't miss the big family enough.

The fact that the Scandinavians rank at the top in the latest global happiness survey, despite their below-2.1 birth rates, shows that perhaps low fertility and happiness can coexist. A society with few babies may still be a happy one.

6. Can politicians fix this? Will the proles vote for lab-grown newborns, collectively ordered and raised by the state?

"What is to be done?" pundits would ask, in good Leninist fashion. If humans don't want babies for themselves and suffer little for that economically and emotionally, can they be persuaded to procreate for a cause larger than their own petty lives?

If persuasion doesn't work, can a democratic government—one "of the people, by the people, for the people"—do something that the people themselves don't want to do? Should we use the government budget to collectively buy artificial-womb-born state-raised replacement humans, Kryptonian style? Would there be a radical reorganization of the human society along the lines of the eusocial beehive model, where a queen-bee state performs the centralized reproduction job for the entire colony? Will common folks vote for this novel use of their tax money?

In case the pragmatic, down-to-earth majority, through democratic votes, considers its own needs worthier than an abstract cause, can a pronatalist dictatorship be imposed forever so that the community can survive its selfish members forever? If that is not sustainable, are we doomed?

6.1 Humans are blissfully oblivious to abstractions like humankind

To inspire people to procreate, pronatalist advocates have tried God (not enough prolific pope-lovers to lift Italian fertility back to replacement) and country (nationalism appeals more to those too old to reproduce), with little success. Given the increasingly global reach of low fertility, they could soon try humankind. Is humanity real and tangible enough to motivate humans where the Almighty and patriotism have failed?

Will people care enough about their species to have more babies? We won't know the answer to these questions until someone engages real humans in the real world, informing them of the unintended consequences of their lifestyles for humanity. It may turn out to be easy; it may be that people do care, and all they need is a gentle wake-up call. Reminded of what is important in life, the peasants would flock to the bedroom, and 40 weeks later, the extinction problem of their species would be fixed for good. If history is any guide, however, procreating for humanity is going to be a tough sell.

The odds are against the salespeople. Of all the previous doctrines urging people to live their lives for an abstraction, not one has passed the test of time. The scripts were the same. According to a charismatic messiah, life is empty without working, breeding, killing, dying for a sacred cause: the people, the class, the race, the faith, and so forth. Yet apart from a few lunatics, the masses never believed strongly, if at all.

"The long and short of it, Fred, is that you've gotten too far ahead of your constituency."

Dedicating one's life to a fanciful abstract concept seems more descriptive of preprogrammed robots, not the animals of flesh and cravings that regular humans are.

Life's worldly temptations were just too strong. Given enough time, history's utopian revolutions all ended up co-opted by a bureaucratic nomenklatura of petty opportunists: the sex-loving popes of the Middle Ages, 20th century's obese communist apparatchiks, and the like. Those were not exactly the transformed and better human beings who should have emerged, in the vision of the first-wave idealists, to ensure the whole enterprise's perpetuity. Somewhat unsurprisingly, most of those oddball idealists would themselves be quickly killed off in the settling down of their revolutions: France's Jacobin leaders, Lenin's associates other than Stalin, and so on. Without fail, all historical utopias quickly turned into corrupted dystopias, run mostly by ruthless characters with base motivations.

Following such traditions, if it was ever imposed, a dictatorship for fertility would not last very long, let alone forever and ever. Given the precedents, a failure to save humankind would once again signal people's chronic intractable apathy toward the abstract, imagined aggregate-level communities, not a lack of saving efforts. Changing people is a thankless business. While advocates of most causes are free to speak these days, their target audiences are also free not to listen. As everyone nowadays is considered equally human and has one similar vote, no one has to listen to your lectures, no matter who you are. If you are less attractive than, say, that hottie on the next channel, you're at the mercy of the remote control.

"YEAH? WELL, MY UNINFORMED VOTE COUNTS JUST AS MUCH AS YOURS, MISTER-GOOD-CITIZEN-NEWS-FREAK."

The species, so real and meaningful to some, may mean nothing to others. Most humans might not think much of humanity, if at all. At best, it only exists as a hollow abstraction for most of its constituents, with much less appeal than their trivially personal pursuits. Which come first and are more real, breathing individuals or the abstract, imagined communities—tribes, nations, species—those humans unwittingly create and incessantly change, just by living? In all nations, USA or USSR, there are always exponentially more souls trying to milk the system than those who believe in that "Ask not what your country can do for you" marching song.

6.2 Mere mortals may care briefly, but how about forever and ever?

Even if the public interest in saving humanity, and the birth rate along with it, could be raised for a while, there would always be the test of time. To keep this 300,000-year-old species around forever, you will have to find ways to maintain people's enthusiasm in, say, 4 million years.

The first step to that, of course, is to make sure your sweetheart won't dump you in four years. Supposing she agrees to give you four beautiful babies, how can you two guarantee that these tiny creatures will, in their days, have enough too? And what about the children of the children of

their children, if some would still be born at all?
The brats you will forgo a lot to raise may, when
they become independent adults, just favor those
forgone activities and ax your favorite species
anyway. Their time will come, and time can make
your world-saving efforts pointless. If humanity
goes away, you know who is fully responsible: the
humans.

"Excuse me, where did you get those sandals?"

If your fellow Homo sapiens don't care,
nobody can stop their species (and it's theirs, too,

Vladimir Ilyich!) from vanishing. Even in a futuristic scenario of state-ordered-and-raised new citizens being conceived in farms of artificial wombs, the final decision on how many to collectively have would still rest with the taxpaying public. Besides the obvious ethical concerns—orphaned children often face greater obstacles in life—would existing humans want to spend precious budget resources on such pooled purchases? No matter how rich our society becomes compared to our ancestors', we will always feel a chronic lack of money at the present moment. The gigantic budget deficits and mountains of debt currently piled up by Earth's wealthiest economies all point to a simple conclusion: money is never enough for humans. Don't pin too much hope, therefore, on a babies-procuring referendum. Average voters might easily argue that every dollar spent on sustaining the abstract species means one dollar less for their actual needs, which are the only things that matter to them.

And even if the majority voted for centralized reproduction for a while, how could we be sure they would still choose that method in 5 million years? Or 5 billion years?

6.3 Leaders gonna lead? Nah, they're just pretentious trend followers

Ultimately, political leaders can't fix these reproductive matters, for they are just our employees. Democratic government is only a purchasing cooperative through which we pool resources to collectively buy public goods: clean streets, safe neighborhoods, well-lighted cities, peaceful borders, and so forth. Glory-seeking politicians are then hired, on a competitive basis, to run the powerful institution. Despite their often-abused day-to-day operational discretion, vote-crazy elected officials know they can be replaced: the electorate fired war heroes Churchill and Bush 41 right after military triumphs. At the end of the day, it's the faceless masses—not the pretentious pols—who decide what "we" want. This explains those stereotype-busting real-world cases of right-wingers building the nanny state in the 1950s and leftists scaling it down in the 1990s. Regarding social change, many smart leaders have consistently proven themselves to be just shrewd trend followers.

Accordingly, poll-obsessed career politicians can't be counted on to fight the birth dearth in liberal democracies. Though aware of how low fertility raises the median age, leaders in the developed world are now concentrating most policy efforts on the healthcare and welfare reforms that accommodate, rather than fix, that root cause of the new aging reality. Where it exists, promoting more childbearing is a tiny government program compared to the massive budget for Social Security, whose task is to feed the coming tsunami of boomer retirees.

On demographic issues, modern Western politics looks centuries away from the pronatalist consensus that prevailed until World War II, with

"According to my latest poll, Sire, it's eleven in favor of continuing the Crusade and eight thousand in favor of returning home."

its funny vocabulary of "selfish women," "superior interests of the nation," "foreign hordes," and so forth. Politicians of such sensibilities are now outside the mainstream of major developed democracies. Understanding the voters' right to decide for themselves, the comrades who make policy today, for their part, prefer to go with—not against—the flow of the people. Weathervane public servants can only serve, not save, a species that consciously doesn't want to save itself.

7. Do intelligent species blindly bloom and die, being just brainier wild nature?

How could an intelligent species of conscious individuals follow a well-known yet inescapable path toward extinction? How is it that humans are aware of this course of events but still unable to avoid it? Is this outcome already coded in our fun-loving genes? Aren't we taught that survival and reproduction are the primary purposes of life itself? Descended from an unbroken line of toughies who survived to reproduce through 3.5 billion years of adversity, are we going to die out from material plenty, of all things? What is happening to our legendary consciousness? And that unique destiny in space?

7.1 An incidental and insignificant humankind in an inexorable macrocosm

All this feels absurd until you stop attributing so much meaning to humanity's existence in the cosmos. It takes a certain degree of insular naivety to see earthlings as unique or destined for something special. Starry-eyed teenagers are rarely told that most of them are destined, in the unfathomably weird plan of the big (wo)man upstairs, to become bland, invisible nobodies. Then we grow up and realize, little by little, the banality of our place in the world. Five hundred years ago, the church imprisoned Galileo for doubting Earth as the center of the whole cosmos. Over the next three centuries, humans gradually discovered other planets in the solar system. Then came other stars, near and far. The star assemblies called galaxies, for their turn, were only firmly determined in the 20th century. The first exoplanet orbiting other solar systems was not pinpointed until 1992, yet thousands more have been located since then.

Will this historical trend toward insignificance lead to the recognition down the road that humanity is just one of countless intelligent species in the universe? Our run-of-the-mill sun is one of about 200 billion stars in this Milky Way galaxy. To go up in scale, there could be, for a middle estimate, 200 billion other galaxies in our observable universe. And other bubble universes/regions in the cosmos might very well exist beyond our cosmological horizon. The probability is such that life must exist somewhere in all this infinite vastness. Or is it everywhere? As noted before, even assuming the super-slim odds of only one such civilization per galaxy, there should still be around 200 billion civilizations across this observable universe.

Nature's gradual, trial-and-error evolution that led to intelligent life on Earth might not be unique to this planet after all. Sentient species may bloom and fade everywhere, blindly and purposelessly, like wildflowers across the cosmic desert. For humans who don't believe in a Creator of worlds and the enactment of his or her premeditated plan, intelligent societies are just a natural phenomenon. As an inspired man once did, let's imagine that heaven doesn't exist and that there is only sky above us. That boundless universe, this little puppy, and the arbitrary

physical constants permitting its existence may just randomly be—not created by a dog-god. Various "He/She generated the world" creation myths are but opium for those souls who cannot handle cold randomness. Your love–hate affair with that gender-fluid imaginary friend named Yahweh is all in your head, having nothing to do with an oblivious physical world that exists on its own. Cosmic civilizations are evolved-yet-ephemeral masterpieces of wild nature. No better than the dinosaurs or the slimy worms, our species didn't choose the when and where of its advent in the universe. And we also might have no say in our later vanishing from this world. Bigger forces of nature are at work here—much bigger than the human mind.

Human consciousness, as impressive as it is, isn't much in the greater scheme of things. This demographic path toward extinction might be one more phenomenon that we are well conscious of but helpless against. Important things in the world, like the whole expanding universe, just run their course, independently of our awareness of or wishes for them. One such thing is the human body, this small piece of meat that hosts each of us. No matter how high up society's food chain you think you are, your soul can neither choose your body's specifics nor prevent its preprogrammed breakdown. Beyond a certain age, your irreplaceable soul case will begin to fall apart. No amount of tofu, yoga, or inner dialogue, prescribed by your keenly aware and worried mind, can stop the decay. The failing body is blind to your spirit's parasitic existence. The self-destruct function keeps running. Soon enough, the organism will have reached its expiry date. Your precious body turns into a dead body that stinks. And you are no more. Happy deathday, you bloody big shot! Hoping to alter the set trajectory of the human species when you can't even escape your own body's preprogrammed fatal course, something you are

"This is a little embarrassing to admit, but everything that happens happens for no real reason."

only too conscious of? Consciousness looks clearly overrated.

7.2 Under a veneer of consciousness, we are those timeless animal instincts

Although humans make a big show of having the most powerful consciousness in nature, we are mostly conscious of what our hearts want. And the human heart is wild. To faithfully describe human beings, the overhyped consciousness is just a sideshow next to the main event: our animal instincts. Intelligent life is nothing but an evolved form of wildlife, highly evolved but no less wild.

Evolution may have made ours the most complex society in all of nature. Yet the fact remains that we still belong to the kingdom Animalia (class Mammalia, order Primates, family Hominidae, a.k.a. great apes), being made of blood, flesh, and cravings. With 98 percent of our DNA shared with chimps, we are merely evolved great apes, our society an evolved ape society. The call of the wild may be subconscious; it is irresistible for these Facebook-addicted primates. It's staying alive and enjoying oneself.

Most of our urges aren't that different from those of other primates, be it playing power-mad politicians (alpha gorillas enjoy asserting their authority over underlings), eating delicacies (chimps hone tools and techniques to extract honey from hives), enjoying intoxication (mandrills are fond of the hallucinogenic iboga root), or just contemplating the world from a hot spring (like those Japanese snow monkeys).

Driven by the same animalistic desires, humans have, over time, become far better than simpler animals at fulfilling these urges by using a distinctly superior brain. Whereas wild creatures spend most of their lives being hungry and prowling for food, obesity is fast becoming one

"I'VE TRIED, MARION, I REALLY HAVE. I WORE THE SUIT,
I'M EATING AT A TABLE IN A RESTAURANT, BUT I
STILL WANT TO CHASE CATS!"

of our major problems. While we still follow the same primal instincts as other great apes—food, sex, pack rankings, and neurological highs—this nerdiest of species has made them much more sophisticated: 24/7 snack vending machines, lap-dance etiquette, Tinder, society balls, final clubs, spirits, music, books, and so on. Even the highbrow highs some humans get by writing terrible concertos or unreadable philosophical treatises still come from the old neurohormones responsible for simian bliss.

An evolved brain doesn't kill our animal desires—it just serves them better.

7.3 Contraceptives reveal how little we want babies versus just sex

While all living things wholeheartedly follow their innate drives, no species—except us—ever forgets to go forth and multiply.

They don't have a real choice. While complex animals are programmed like robots to seek sex, no one else besides humans can control the reproductive consequences. Nature gives our animal siblings a lousy deal; going in for some quick fun, they end up with something they have no use for: babies. Bears or boars aren't after old-age support or Social Security preservation when mating; most aged creatures die alone and abandoned. Yet sexual desire in the woods frequently leads to unplanned offspring. Babies in the animal world are all unintended accidents—the uncontrollable byproduct of casual sex.

Brainy humans, in contrast, have succeeded in separating the two intertwined sides of sex—recreational and reproductive—becoming thereby the only life-form that can have the fun without the function. Invented by nature's most evolved brain, effective contraceptives have

"Sometimes I wonder if there's more to life than unlocking the mysteries of the universe."

89

started a brand-new chapter in natural history. For the first time, an animal can now override the hard-coded directive that has made sure, from time immemorial, it will blindly multiply. Sex, which most of us are programmed to enjoy, no longer automatically means a multitude of babies—something many humans might not enjoy as much.

With contraceptives, we can now identify what percentage of human sex is just for fun and how truly minuscule the percentage for reproductive sex really is. By permitting sex with little chance of pregnancy, contraceptives reveal that the basic instinct for babies (if such an innate desire actually exists) is much weaker than the one for casual sex. Having babies today is often a deliberate decision, taken in full consciousness and intent. Reproductive sex is rare and awkward next to the mountains of recreational sex those wild Homo sapiens are having.

As pills and condoms open the technical path to low fertility, our now-unshackled animal hearts will lead us forward. What we do with this new contraceptive-afforded liberty is to double down on the eternal instinct for fleshly fun, which is part of human nature itself. We are still not very far beyond these instincts.

7.4 Evolution conceptually ends when a species purposely stops reproduction

Evolution might be revealing its conceptual endpoint, where the brainiest species can and does party itself out of existence, consciously and happily. In doing so, the most evolved animal will invalidate the first assumption of evolution: life always tries to reproduce itself into the future. That is not correct anymore in the case of these super-brainy wild creatures called Homo sapiens. This latest chapter of low fertility and demographic decline could very well be the last one in the evolution book as evolution finally offers a species the unprecedented possibility for self-elimination.

The natural road to self-extinction is opened by evolution itself. Coming from evolved primate brains, liberating contraceptives and helpful robots are as much products of natural selection as the chimpanzees' sharpened spears or the beavers' dams. When a species evolves enough brainpower to survive the kind of externally forced deadly hardships typical of lower-species

91

extinctions, it will naturally self-terminate in pursuit of fun. Smart or dumb, no species will escape nature's birth–death law.

Even if an intelligent society evolves on Earth after humanity, if its members also eat/drink/advocate/philosophize/enjoy themselves into oblivion, then we may already be experiencing the theoretical limit of biological evolution. When served by too much brainpower, basic animal instincts can lead to a willing death-by-fun

for the terminally evolved species. To put this in Buddhist terms, those species that successfully escape the endless karmic cycle of ignorance and suffering will attain a nirvana of ... ahem ... nothingness.

7.5 Humanity being conscious of its own fate will not stop the process

When humans finally realize the unintended consequences for the species of their new reproductive preferences, will the awareness of demographic doom prevail over that of their self-centered cravings? Can people deliberately prioritize the abstraction of humanity over their worldly desires? It's hard to do so when those visceral urges primarily define who one is. "This is me, wild at heart. Only my primeval desires are genuine, not that abstract, derivative aggregate called humankind."

Even the highbrow intellectuals, theoretically the most aware members of our great-ape society, are not above their little hormonal pleasures. Our learned and well-informed brothers and sisters are as active as anyone else in ruining humankind's future, having traded that additional baby for the dopamine-based highs of knowledge, poetry, philosophy, and so forth.

A series of intentional choices dictated by basic instincts, all putting the corporeal individual above the abstract society, will guide humanity down this road of conscious self-extinction. First, we mindfully opt for the former when faced with the trade-off between various nonreproductive hobbies and humankind's numerical maintenance. Next, adult citizens will explicitly vote to spend their government's budget on themselves, not on lab-born, state-raised replacement humans, which is a technically feasible way out for humanity. Demographic aging and decline will proceed every step of the way, regardless of humankind's complete awareness of those steps.

At the end of the day, the conscious mind doesn't do much beyond taking note of what our wild hearts desire. While some hearts are in love with the lofty, abstract concept of humanity, most may feel nothing at all. Next to the tyrannical animal instincts, consciousness looks like a meek and inconsequential clerk. We are fully conscious of, but still powerless against,

Victoria Roberts

"You're born, you deconstruct your childhood, and then you die."

nature's major preset courses: personal dying, demographic aging, and cosmic expansion.

7.6 Do aliens also age and die?

Does the same aging fate apply to other intelligent societies across the universe? Can humanity be truly alone in infinity when Earth's astronomers estimate that there are 100 million Earth-like planets in just our Milky Way galaxy, one of hundreds of billions of galaxies out there? How have those faraway cosmic civilizations evolved? Do they follow the classic cycle of birth and death? Have some of them figured out how to survive the coming physical demise of the entire universe? If they are out there, why have they never contacted us?

If the hallmark of science is the need for empirical evidence to substantiate theoretical claims, then to answer these questions scientifically, we must start with the only real-world civilization that's currently researchable: ours. Based on our super-aged societies here on Earth, could we extrapolate population aging and decline as the natural ending common to all intelligent species in the universe—past, present, and future?

We may never know for sure. Aliens could be so different that none of our concepts would apply to them. ETs would be genuinely different from us if they were immortal, having vanquished the Grim Reaper once and for all. Such an endless existence is mathematically impossible, however, because the physical universe itself is not eternal. A being cannot exist before the universe's birth 13.7 billion years ago or continue to live after its ultimate destruction. Well, except for God herself, obviously.

Cosmic demise aside, immortality is also not possible due to the usual accidents of life. Statistically, your probability of having a fatal mishap, like a traffic collision or a fall down the stairs, rises over time and climbs to almost 100 percent after just a few thousand years. Consequently, as with some jellyfish, hydras, and other aging-free primitive life-forms, a hypothetical, biologically immortal human will still die eventually—on average, after 500 to 600 years—because of a deadly accident.

Elsewhere across the universe, aging-free gooey aliens that reproduce asexually, if such a combination exists, will still succumb to the general hazards of life, facing, over time, an increasingly bigger risk of death until they are fatally hit by one of the dangers. For both individuals and species, there's always Murphy's

"Those once-in-a-lifetime events are beginning to add up."

Law when it comes to the precise causes of death: whatever can go wrong will eventually go wrong.

Without individual immortality, all intelligent beings in the universe should understand dying. Getting psychologically ever closer to death should be a truly universal experience, something sentient beings billions of light-years apart can all intimately relate to. Like your home star, you were born, you live, and then you die. You can be around for 100,000 Earth/Pluto/Krypton/ whatever years; all you have is your being in this moment and the knowledge that your existence is not forever. There is a before-you and an after-you. All lives—you, me, the winged ETs 10 billion light-years away—are indeed equal before time. In this sense, death may unify all creatures of the universe.

If also composed of mere mortals, then all ET societies should have a measurable median age. And that median age should rise when existing members of those alien societies prioritize other hobbies over creating, sexually or asexually, new members to replace themselves. Barring youthful accident or suicide, intelligent species will die of old age just like individuals.

7.7 Are ETs not visiting because they're just jaded oldies?

Similar to individual organisms, all cosmic civilizations will be equal before time and death, being nothing but ephemeral and purposeless masterpieces of blind nature. Like wildflowers in the lost oases of the great cosmic desert, civilizations are born, age, and die. To some extent, our demogra-fate theory might have already provided an answer to the famous Fermi paradox of why ETs don't contact us if they do exist.

Absent external survival threats, the more senile and jaded a society gets, the less passionate about space conquests it may become. If we look at our aging society, we can make educated guesses about the aliens' much older worlds. Just as tons of tired old couch potatoes of our own species are passive about venturing far from their living rooms, why should the high-median-age aliens be any different? Perhaps the same aging process has made our technologically advanced neighbors prefer quiet tai chi to Hollywood-style space conquests? Old age will spare no one, not even those fearsome, world-wrecking aliens.

Why does the Fermi paradox so confidently assume that ETs want to meet us? Is it because those imagined aliens are merely a projection of ourselves—a still young, hungry, and restless bunch? Why can't we envision old and satiated ETs who don't bother to get up, having lost the thirst for adventure? Sci-fi fantasies about rapacious colonialist invaders from the sky may unwittingly reveal more about our youthful and barbaric recent past than about the aliens from the future.

The extraterrestrial conquerors of worlds in Hollywood blockbusters strangely remind us of brutal conquistadors like Hernán Cortés—the

"Harold and I were staying home long before it was fashionable."

young and greedy 16th-century alien from Europe who landed in Central America and destroyed the technologically inferior Aztec civilization. It feels like yesterday when rapidly aging Western societies were still demographically young and getting all hormonal over collecting faraway dominions. Those cruel and vainglorious colonial empires were not dismantled until the 1950s, a mere 70 years ago.

Passions have calmed down a lot since those wild days. As nations get older and more affluent, humans are quietly maturing out of their barbarian ways. Technological progress barely gives parts of humanity a century of prosperity, and somehow many countries have reconsidered the wisdom of reaching out to uncontacted indigenous tribes. Not-so-rich India decided, for example, to respect the North Sentinel Islanders' wish to remain in the Stone Age. Is humankind itself an uncontacted tribe of the universe, considerately left alone on our little green island by much more enlightened civilizations?

Regarding space colonization, as soon as we could crawl to the moon in the 1960s, humans, increasingly richer and older, quickly became lukewarm about both space and multiplying into it. Space-related spending as a percentage of the overall budget has dropped in most developed societies. Further down the road, after millennia of good high-tech life and with age-related wisdom, spacefaring civilizations that reach the technological stage of intergalactic travel might completely lose interest in exploring the cosmos or finding aliens.

THE MILKY WAY
(Detail)

Why do insular humans think their dime-a-dozen species is so exceptional that ETs feel compelled to visit? Why would a millions-of-years-old civilization want to cross the universe to meet this 300,000-year-young species if the former has become like Grandpa, who has seen enough and now just wants to quietly enjoy TV? A general awareness of other species' existence out there could be enough for those sweet old-timers. Their super-aged societies might be too busy enjoying a peaceful, fun-filled demographic atrophy. For that, could it be that there may be no place like home?

How could anyone conquer the universe anyway? It's infinity up there. The very desire to travel that endless void of space is a sign of youth, a time when not even the sky is the limit. Then you get older, and those sweet, heady dreams fade. As the saying goes, "Where kids see the beach under the pavement, the old see the pavement under the beach." The mellow civilizations of the universe may just prefer the everyday joys of home, seeing nothing but pavement beyond their great purple/maroon/lime/blue yonder.

Interstellar migration may look like a pointless merry-go-round to those world-weary eyes. Why bother going to other worlds if all you see is the same cycle everywhere? Going to the stars sounds terrific, but what about the aftermath? Without a disaster on your home planet, for what reason do you resettle to a faraway world? To resume, once there, what is happening here and now: the moment folks begin to feel comfortable and take survival for granted, they won't even care to reproduce enough to maintain their society? This whole space-colonization thing sure looks like one hell of a crazy hamster-wheel ride.

"THE CONCEPT IS SIMPLE. PEOPLE CALL IN WITH QUESTIONS ON EXISTENCE AND REALITY, AND YOU RESPOND WITH TOTAL SILENCE."

7.8 As part of nature, wild species like humanity just bloom and fade

A birth–death life cycle for intelligent species is just the way of nature. Complex animal societies constantly rise and fall around us. Extinction is as old as life itself. Today's living kinds may account for less than 1 percent of all the life-forms that have ever roamed Earth. Nature is in constant renewal; billions of species have died out, only to be replaced by newer ones. After humans depart, the eternal jungle will absorb the remnants of our technological civilization so completely that in 10 million years, virtually no traces will be left of its previous existence. No future network-connected octopus, cockroach, or centipede will have us, those flamboyant great apes, in their memory.

Intelligent life is still life, forever wild at heart. Fundamentally, intelligent life-forms are just another part of wild nature, evolving and fading namelessly and innumerably across the cosmos. Human/alien societies are but evolved animal societies, with better beaver dams and deadlier hunting techniques. It's a cosmic desert out there—endless, hostile, but sporadically dotted with lush oases of life, full of births, deaths, and births again. There might have been billions of terminally evolved sentient species in the universe's 13.7-billion-year history. Perhaps only a tiny fraction of those species are still alive—

most may have gone through their life cycles, yet endlessly more civilizations are being and will be born, magically springing up like ephemeral wildflowers. Although intelligent life may have arisen separately in lonely oases, its wild nature and fleeting destiny are the same across the cosmic desert.

Evolution's conceptual end point is finally reached when the brainiest wild species can and does party itself out of existence, consciously and merrily. Countless waves of evolved life have headed and will head for this eternal shore, billions of years ago or billions of years from now. When these app-addicted apes become

Yeah, sure

fossils in this lost oasis, somewhere in a nearby galaxy, another wild species will randomly evolve to boost brainpower, create technology, solve hunger, enjoy fun, reproduce less . . . This ancient cycle of blind evolution may have played out a zillion times across infinity.

Given no say whatsoever in their wild birth, wild-at-heart cosmic civilizations also have no veto on their natural death. There's no good, bad, or purpose to intelligent life anywhere—it's all just wild nature, evolved to be a bit less dopey perhaps.

This epic evolutionary spectacle of humanity runs its own course over millions of years, and you—a fluky nanosecond of consciousness and curiosity—can do nothing but play tourist and enjoy your part in the action. Your dear species having an inexorable life cycle is hardcore fatalism, but so is your ephemeral alien body dying against the mind's wish. Make peace with the relentless, unfeeling nature in and around you. Face up to your fleeting moment.

"Good Lord, Professor! There it goes—the theory of everything!"

Wikipedia References

I happen to think less is sometimes really more. For this book, I purposely chose to reference Wikipedia articles instead of academic journals because they are convenient and accessible starting points for more readers. I also did not include footnotes to keep the book short and simple.

If you are interested in learning more about a topic (e.g., Ageing of Europe), you can use Wikipedia's search box to quickly access an article instead of typing the whole URL.

https://en.wikipedia.org/wiki/List_of_sovereign_states_and_dependencies_by_total_fertility_rate

https://en.wikipedia.org/wiki/Sub-replacement_fertility

https://en.wikipedia.org/wiki/Aging_of_Japan

https://en.wikipedia.org/wiki/Ageing_of_Europe

https://en.wikipedia.org/wiki/Estimates_of_historical_world_population

https://en.wikipedia.org/wiki/Projections_of_population_growth

https://en.wikipedia.org/wiki/Timeline_of_the_far_future

https://en.wikipedia.org/wiki/Chronology_of_the_universe

https://en.wikipedia.org/wiki/Ultimate_fate_of_the_universe

https://en.wikipedia.org/wiki/Exoplanet

https://en.wikipedia.org/wiki/Planetary_habitability

https://en.wikipedia.org/wiki/Abiogenesis

https://en.wikipedia.org/wiki/Astrobiology

https://en.wikipedia.org/wiki/Biological_immortality

https://en.wikipedia.org/wiki/Anthropic_principle

https://en.wikipedia.org/wiki/Fine-tuned_universe

https://en.wikipedia.org/wiki/Why_there_is_anything_at_all

Disclaimer and Acknowledgment

Although the publisher and the author have made every effort to ensure that the information in this book was correct at press time and while this publication is designed to provide accurate information in regard to the subject matter covered, the publisher and the author assume no responsibility for errors, inaccuracies, omissions, or any other inconsistencies herein and hereby disclaim any liability to any party for any loss, damage, or disruption caused by errors or omissions, whether such errors or omissions result from negligence, accident, or any other cause.

All cartoons used in this book are lawfully sourced from www.cartoonstock.com.

9 798531 458612